HOW THINGS WORK!

ELECTRICAL GADGETS

ADE DEANE-PRATT

WAYLAND

First published in 2011 by Wayland

Copyright © Wayland 2011

Wayland
338 Euston Road
London NW1 3BH

Wayland Australia
Level 17/207 Kent Street
Sydney NSW 2000

Editor: Julia Adams

Produced by Tall Tree Ltd
Editors, Tall Tree: Rob Colson, Jennifer Sanderson
Consultant: Penny Johnson
Step-by-step photography: Ed Simkins, Caroline Watson
Designer: Jonathan Vipond

British Library Cataloguing in Publication Data
Deane-Pratt, Ade.
  Electrical gadgets. -- (How things work)
  1. Electric apparatus and appliances--
    Juvenile literature.
  I. Title II. Series
  621.3'1042-dc22

  ISBN-13: 9780750261258

Printed in China

Wayland is a division of Hachette Children's Books,
an Hachette UK company.
www.hachette.co.uk

Picture credits

The publisher would would like to thank the following for their kind permission to reproduce their photographs: key: (t) top; (c) centre; (b) bottom; (l) left; (r) right 1 and 9 (br) Derek Latta/-Stock; 2 and 17 c Kosmonaut/shutterstock.com; 4 Traffictax/Dreamstime.com; 5 and 32 Snr/Dreamstime.com; 7 (tr) Alexandrg/Dreamstime.com, (b) Wavebreakmedia Ltd/ Dreamstime.com; 8 Restyler/Dreamstime.com; 9 (tl) Daleen Loest/Dreamstime.com, (bl) Feblacal/Dreamstime.com; 11 (tr) Mstephan/ Dreamstime.com, (bl) Stuart Key/Dreamstime.com, (br) Joris Van Den Heuvel/Dreamstime.com; 13 (tl) Julia Nichols/iStock images, (tr) BMW AG, (b) Photographerlondon/Dreamstime.com; 15 (tr) Wayland, (bl) Pedro Antonio Salaverría Calahorra/Dreamstime.com, (bl) Orest/Dreamstime.com; (tl) Robgooch/Dreamstime.com, (bl) Yunus Arakon/iStock; 19 (tl) Otnaydur/Dreamstime.com, (tr) Greg Nicholas/iStock, Paulglover/Dreamstime.com; 21 (tr) Andrew Howe/iStock, (c) JYI/shutterstock.com, (bl) Dary423/Dreamstime.com

The website addresses (URLs) included in this book were valid at the time of going to press. However, because of the nature of the Internet, it is possible that some addresses may have changed, or sites may have changed or closed down since publication. While the author and publisher regret any inconvenience this may cause the readers, no responsibility for any such changes can be accepted by either the author or the publisher.

Disclaimer

In preparation of this book, all due care has been exercised with regard to the advice, activities and techniques depicted. The publishers regret that they can accept no liability for any loss or injury sustained. When learning a new activity, it is important to get expert tuition and to follow a manufacturer's instructions.

CONTENTS

POWERED UP

Every day, we rely on gadgets to make our lives easier. From alarms to telephones, these gadgets are powered by electricity.

A SOURCE OF POWER

Large machines, such as televisions, use electricity from the mains. This electricity is generated at power stations and sent to our homes along power lines. Many smaller gadgets, such as remote controls, are powered by batteries. Batteries contain chemicals that produce electricity. Batteries have to be replaced or recharged when they run flat.

Lightning is a massive flow of electricity between storm clouds and the ground. This is called static electricity and it is the same kind of electricity that makes your jumper crackle.

MAKING ELECTRICITY WORK

The technology behind most gadgets uses electrical components, such as switches and resistors. This book looks at some of these components and how they work in gadgets, such as touchscreens, games consoles and batteries. The project pages will show you how to make your own gadgets, such as a fan and a circuit game.

Small, portable gadgets are powered by batteries. This mobile phone uses electricity from a battery to send and receive messages and to show the display on the screen.

WARNING! Electricity is dangerous so safety is very important! Never play with electrical appliances or wires. Make sure you switch off the power at the socket first before unplugging or plugging in gadgets.

Large gadgets and those that do not need to be moved around are powered by mains electricity. These include microwave ovens, televisions and desktop computers.

HOW DOES IT WORK?

The mains electricity found in our homes is generated in power stations. Electricity at high voltages is sent from the power stations to substations, which make the voltage lower for us to use safely in our homes. Mains-powered gadgets, such as electric kettles and televisions, need to be plugged into a socket using a lead that ends in a plug. The socket is attached to a network of wires that carry the electricity from the substation. A switch on the socket turns the electricity supply on or off.

Try it !

Go into each room in your home and find the sockets. How many gadgets do you have permanently plugged in? How many of these gadgets do you use every day?

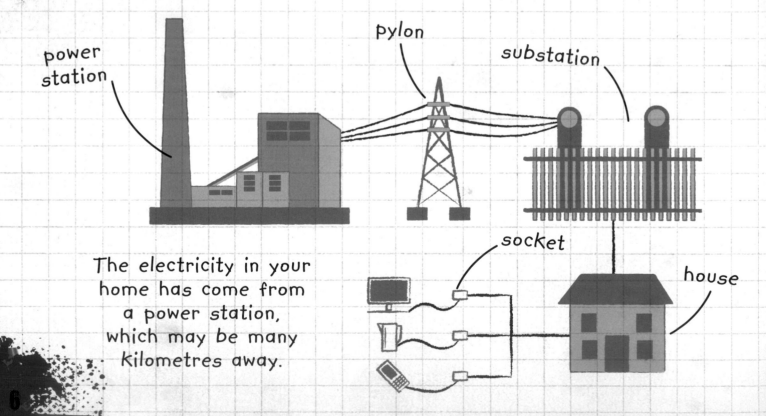

power station

pylon

substation

socket

house

The electricity in your home has come from a power station, which may be many kilometres away.

MAINS IN ACTION

While televisions, computers and games consoles run off mains electricity, we can plug additional devices into them. These devices use power from the original gadget's supply to work.

The keyboard and mouse on some computers are plugged into the computer itself. They are powered by electricity that has passed from the socket through the computer. Other gadgets, such as speakers, may also be powered through a computer.

Large flat-screen televisions use a lot more electricity than old-fashioned televisions. One of the reasons for this is that the image flat-screen televisions produce is more detailed.

PORTABLE POWER

Batteries are found inside all sorts of small gadgets, such as hand-held games or mobile telephones.

HOW DOES IT WORK?

Batteries provide small amounts of electricity. Chemicals inside a battery produce electricity when it is connected to a circuit. However, batteries do not last forever and need replacing when they go flat.

To avoid replacing flat batteries, people use rechargeable batteries. When these batteries are recharged, the chemical reaction that made the electricity is reversed. Rechargeable batteries can be used over and over again.

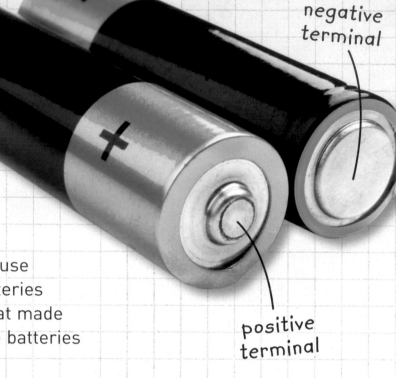

negative terminal

positive terminal

Try it !

Open the battery compartment of a small electric gadget that contains two batteries, such as a remote control or a torch. Take out one of the batteries and put it back in the wrong way round. Does the gadget still work?

When batteries are connected to a circuit, electricity flows between the positive terminal and the negative terminal.

WARNING! Never cut or break open a battery. The chemicals inside are harmful.

The remote control of this toy uses battery power. It sends a signal to the helicopter, which has a battery-powered motor to make it fly.

BATTERIES IN ACTION

Some gadgets, such as remote controlled toys, need more electricity than one battery can provide, so they use two or more batteries. Devices that use a lot of battery power, such as laptop computers, often use rechargeable batteries, which are more environmentally friendly.

Some batteries are very small and thin to fit inside small gadgets, such as this watch. Smaller batteries make less electricity than larger ones, but a watch needs only a very small amount of electricity to work.

battery

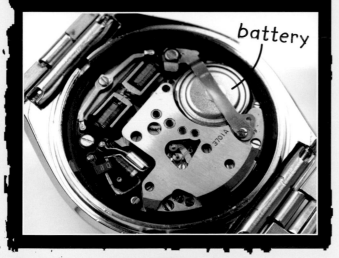

MP3 players use rechargeable batteries to save you having to change the batteries. The device plugs into the mains to recharge the battery or it can be recharged as the MP3 player sits in its dock.

SMART CIRCUITS

Electricity flows only when there is an unbroken loop of wire for it to flow around. The loop is called a circuit. Gadgets, such as touchscreens, work by completing or breaking circuits.

HOW DOES IT WORK?

A simple circuit carries electricity from the power source, such as the battery, to the gadget and back to the power source. When the circuit is broken, the electrical current stops. When you press a touchscreen, you push together two thin layers in the surface of the screen, completing a circuit. This tells a computer where on the screen you have touched.

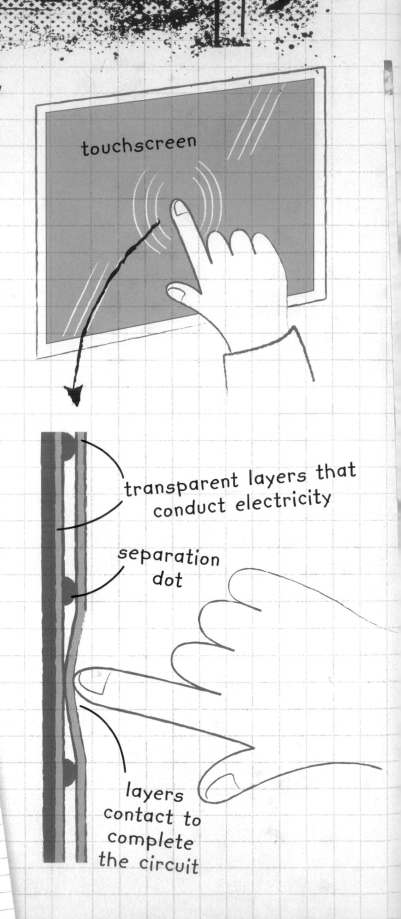

touchscreen

transparent layers that conduct electricity

separation dot

layers contact to complete the circuit

Try it !

Make a simple circuit by connecting a light bulb to a battery. What happens if you disconnect the wire at one of the terminals of the battery?

CIRCUITS IN ACTION

Some circuits, such as the one that connects a mains socket to the power station, may be many kilometres long. Other circuits, such as those found on a circuit board, are just a couple of millimetres across. A circuit board may contain hundreds of interconnected circuits.

This gaming board works like a touchscreen. When players stand on the board, they close the circuit and allow the electricity to flow through the device.

Inside a mobile phone, there is a circuit board. When you press a button, the circuit board transfers signals from under the buttons to different parts of the phone.

Microchips are small enough to fit on your fingertip. Each one contains an electric circuit that may have millions of parts to it. Powerful computers contain many thousands of microchips.

SWITCHING IT ON

Switches turn an electrical current on or off, or change the path of the current. Gadgets can use switches to change speed, direction or power.

HOW DOES IT WORK?

The simplest switches turn an electrical current on or off. When the switch is in the 'on' position, the circuit is complete and the current can flow. When the switch is turned to 'off', there is a gap in the circuit and the current stops. In more complicated gadgets, switches are used to change the path the current takes. For instance, a microchip may have millions of switches on it, so that electricity can flow along different paths.

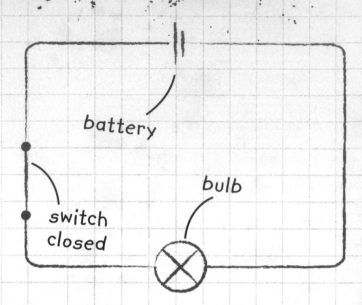

Circuit diagrams use different symbols to represent each part. The circuit above has a switch in the on, or 'closed', position. The circuit below has a switch in the off, or 'open', position.

Try it !

Make a simple circuit using batteries, wires, a light bulb and a switch, laying them out as in the diagrams on this page. When the switch closes the circuit, electric current will flow and turn the light on. Try to turn the light on and then off again.

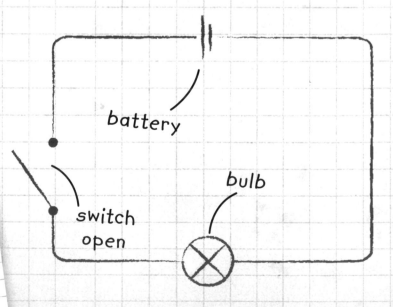

A timer switch can be used to operate lights. The switch turns the lights on and then off at the times set by adjusting the dial.

The switches on the steering wheel of a Formula One car allow the driver to control different parts of the car, such as how much fuel reaches the engine.

SWITCHES IN ACTION

Different gadgets use various types of switch. Most have a simple on/off switch, while others, such as a food processor, have switches that change the speed of movement. The switch controls the amount of power that turns the gadget's motor. The more power, the faster the motor turns.

When the man filming with this digital video camera presses the 'record' button, he is activating a switch that directs electricity to the part of the circuit needed to record moving images.

RESISTING THE FLOW

Inside all electrical gadgets are components called resistors. Resistors control the amount of electricity that flows through the gadget.

HOW DOES IT WORK?

As its name suggests, a resistor resists the flow of electricity through a gadget. The higher the resistance, the smaller the current that can flow. Each gadget needs a particular current to flow through it to work properly. It has a resistor of just the right strength to produce the correct current.

battery

resistor

bulb

Try it !

Take the simple circuit you made earlier and add an extra bulb. The light bulb has a resistance. What happens to the brightness of the bulbs now that there is more resistance in the circuit?

In a circuit diagram, resistors are shown as rectangles. Gadgets such as light bulbs also act as resistors. They allow enough current to flow to make them glow, but give enough resistance to stop them from melting.

RESISTORS IN ACTION

The flow of electricity through a gadget can be precisely controlled using combinations of resistors. Variable resistors change the current in volume or brightness controls on gadgets such as televisions or MP3 players.

Variable controls on music decks, such as volume and bass, work by changing their resistance to the electrical current. The greater the resistance, the smaller the current and the lower the music's volume.

By turning the dial on the iron, you control a resistor. The greater the current through the circuit, the hotter the iron becomes.

Circuit boards inside computers use resistors to control the flow of power across them. If the resistors overheat, the circuit fails, so a fan blows air across it to keep it cool.

A range of different lights are used in electrical gadgets. Some appliances use liquid crystals or lasers for their light, but many use a small light called an LED.

HOW DOES IT WORK?

Light Emitting Diodes (LEDs) act like tiny light bulbs. They are found in all kinds of devices: they form the display on digital clocks, transmit information from remote controls, light up watches, and tell you when your gadgets are turned on. Unlike incandescent bulbs, they do not have a filament that will burn out and they do not get especially hot. Instead, LEDs contain special materials called semi-conductors, which produce light when electricity passes through them.

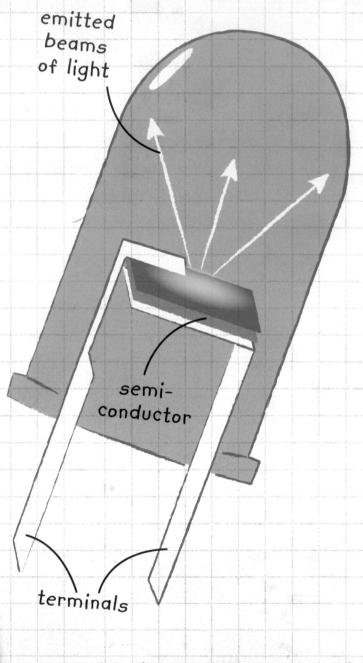

emitted beams of light

semi-conductor

terminals

LED (Light Emitting Diode)

Try it !

Make the same simple circuit you made on page 12, but use an LED instead of the incandescent light bulb. Which light was brighter? What happens if you add more LEDs to the circuit?

7:00

The display on some alarm clocks is made of LEDs. The lights are arranged in a pattern that looks like a digital eight. Depending on the time, different LEDs light up.

LIGHTS IN ACTION

The face of some alarm clocks, digital radios and DVD players use LEDs to light them. In gadgets such as e-book readers and televisions, LCDs are used to light up their screens.

A laser pointer shines a beam of coloured light, which lights up a small dot on any surface it hits. Laser pointers are used in presentations for showing the audience specific things.

The screen of an e-book reader is lit up by an LCD (Liquid Crystal Display). LCD screens have two transparent panels with a liquid crystal solution between them. Light shines from behind the panels and each crystal either allows light to pass through or blocks it. The arrangement of the crystals forms the image.

MOTORS AND GENERATORS

Many machines are powered by electric motors, which turn electricity into movement. Generators work like motors in reverse, turning movement into electricity.

HOW DOES IT WORK?

Electric motors use a combination of electricity and a magnet to make a turning motion. When current flows through a wire that surrounds an iron core, it makes the wire act like a magnet. This temporary magnet – called an electromagnet – pushes and pulls against a permanent magnet placed near the wire, and this creates turning motion.

Try it !

Have a look at a racing car set, electric train set or other similar electric toy. Can you see how the electric motor connects to the wheels in order to get them moving?

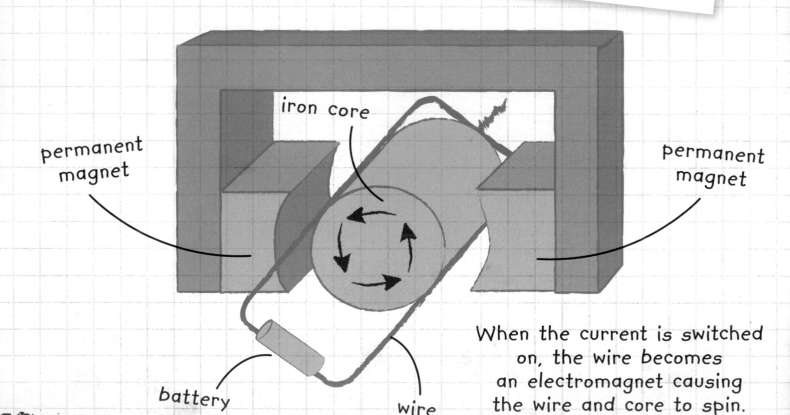

iron core

permanent magnet

permanent magnet

battery

wire

When the current is switched on, the wire becomes an electromagnet causing the wire and core to spin. When the current is switched off, the spinning stops.

MOTORS AND GENERATORS IN ACTION

Electric motors are found in all kinds of gadgets, from huge motors in cranes to tiny ones inside computers. Power stations use generators to make electricity. Smaller generators are found in wind-up gadgets.

motor

laser

The motor in a fan turns the blades, creating a draught of air. If a higher fan speed is needed, a variable resistor allows more current to flow to the motor, making it spin faster.

A small motor inside a DVD player spins the disc so that a laser reader can read the information on it.

This torch-radio is powered by a rechargeable battery. The battery is charged using a generator, which turns the winding motion of the handle into electricity.

STATIC ELECTRICITY

Sometimes electricity does not flow in a current. Instead, the electricity builds up a charge. This is called static electricity, and it is used in gadgets such as photocopiers. Sometimes the charge may get high enough to cause a spark.

HOW DOES IT WORK?

For electricity to move between two points, each point needs to have an electrical charge. One point is positively charged, and the other is negatively charged. The two charged points are attracted to each other. Inside a photocopier, a drum covered in a pattern of positively charged toner rolls over a piece of negatively charged paper. The toner sticks to the paper as the opposite charges attract.

Try it !

Sprinkle some ground pepper on the bottom of a plastic tub. Put the lid on the tub, then rub the tub with a woollen scarf or sweater. This creates static electricity. What happens to the pepper? What happens when you stop rubbing the lid?

paper has negative charge

drum with pattern of positively charged toner rolls as paper moves underneath it

toner sticks to paper

photocopier

STATIC ELECTRICITY IN ACTION

During a lightning strike, enough electricity passes from the cloud to the ground to power a small town for a year! Tiny amounts of static electricity can build up in clothes in a washing machine, making them crackle when you put them on.

As clothes are tossed about in a tumble dryer, static electricity builds up in the fabric. To reduce this, you can add special dryer balls that lift and separate the clothes as they tumble.

People who work with electrical gadgets often wear antistatic wrist straps. These 'bracelets' are connected to the gadget and stop static electricity building up. A spark caused by static electricity could damage the gadget.

A photocopier scans the page to be copied using bright light. The light makes a pattern of electric charge in the toner on the drum.

21

Make a circuit game

You will need a steady hand to play this game. One twitch and you will complete the circuit and lose!

what you need

- pencil
- shoe box with lid
- metal coat hanger
- modelling clay
- LED
- scissors
- insulated wire
- drinking straw
- batteries in a battery holder
- sticky tape

1 Mark three points of a triangle on the shoebox lid. Use the pencil to pierce through the holes.

2 Ask an adult to undo the coat hanger and bend it into an interesting shape. Using modelling clay, fix one end of the wire to the lid at one of the holes. Leave the other end of the coat hanger unattached until step 7.

3 Use scissors to remove about 2.5 cm (1 in) of the insulating plastic from each end of a 30 cm (1 ft) length of wire. Wind one end of the wire around the piece of coat hanger sticking through the lid. Attach the other end to the LED.

4 Cut a 30 cm (1 ft) length of insulated wire and strip 10 cm (4 in) of the insulation from one end. Bend the stripped 10 cm of wire into a loop and twist it around the base of the loop.

5 Feed the insulated wire through the drinking straw and fasten it with tape to make a handle.

6 Slip the loop onto the coat hanger then feed it through the third hole in the lid. Attach the loop to the batteries. Attach the batteries to the LED with a third short piece of insulated wire. Pierce a hole in the side of the box and push the LED through it.

7 Secure the loose end of the coat hanger to the lid with some modelling clay. Try to pass the loop along the wire without lighting up the LED.

Take it further

- Make the game harder by bending the coat hanger so that there are more twists and turns.
- Change the LED for a buzzer so that it makes a sound every time you touch the hanger.

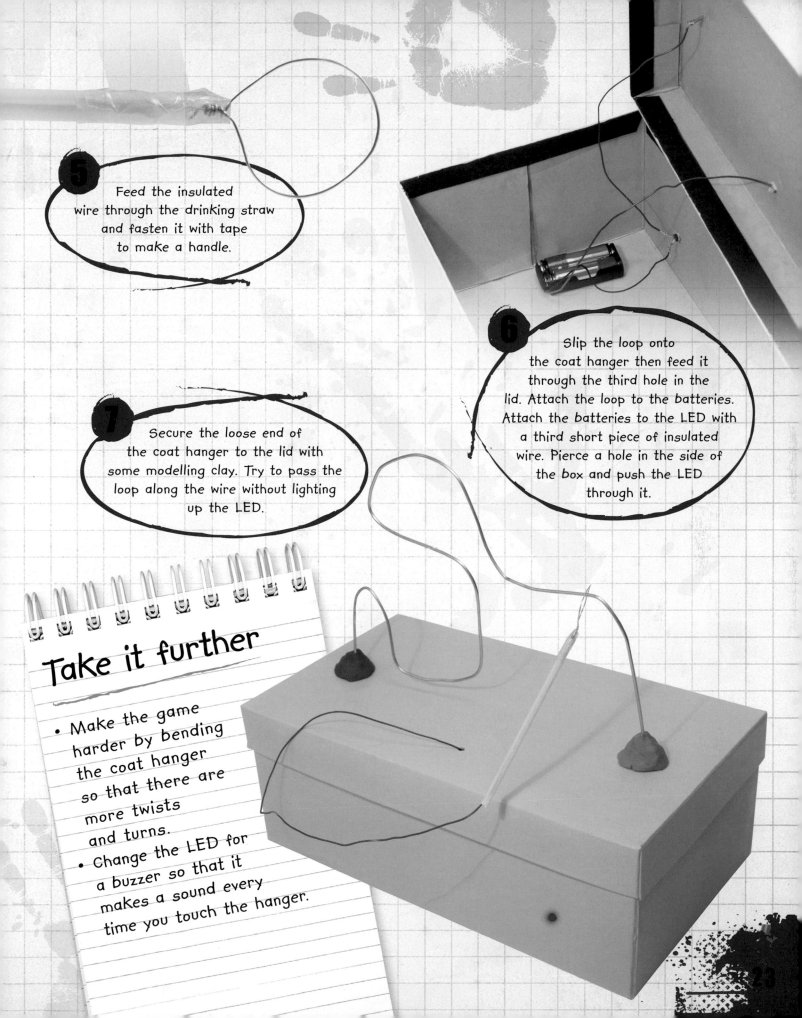

Make a potato battery

Potatoes contain chemical energy, which you can turn into electricity to power an LED.

what you need

- 2 zinc-coated (galvanised) nails
- 2 potatoes
- 2 copper coins
- wire cutters
- 3 10-cm (4-in) lengths of thin insulated wire
- 6 small crocodile clips
- LED

1 Push one of the nails into one side of a potato. Push one of the coins into the other side of the potato. Do the same to the other potato.

2 Use the wire cutters to strip the insulation from both ends of each piece of wire. Attach a crocodile clip to each end of each wire.

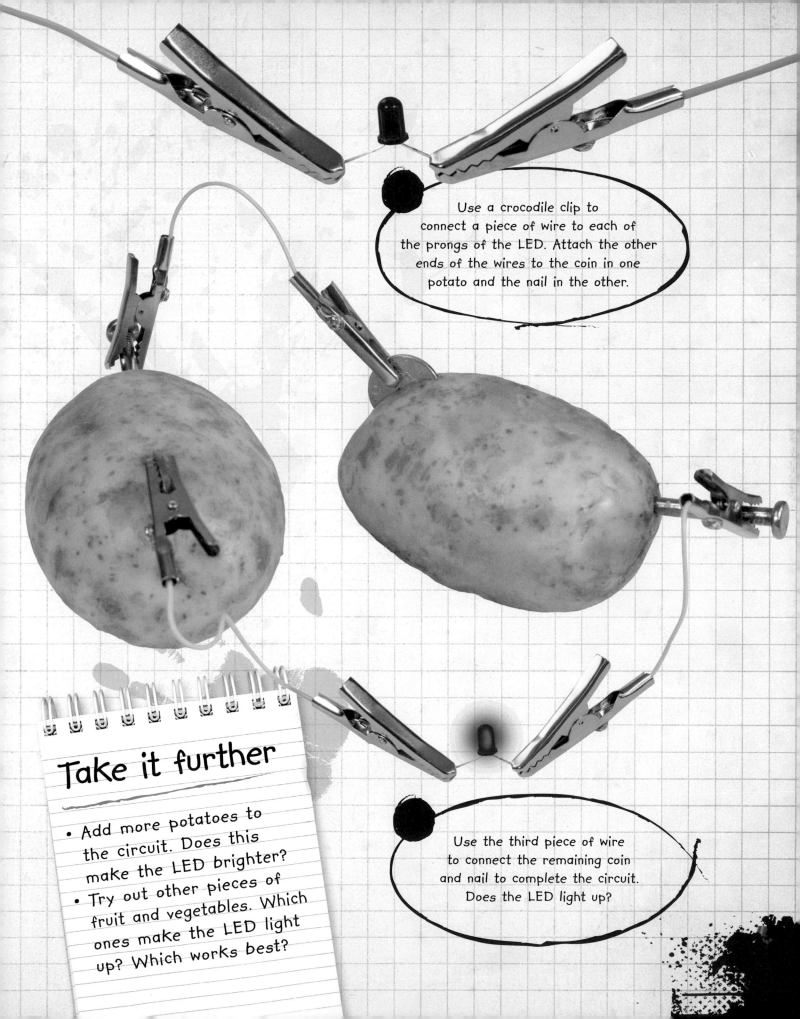

Use a crocodile clip to connect a piece of wire to each of the prongs of the LED. Attach the other ends of the wires to the coin in one potato and the nail in the other.

Take it further

- Add more potatoes to the circuit. Does this make the LED brighter?
- Try out other pieces of fruit and vegetables. Which ones make the LED light up? Which works best?

Use the third piece of wire to connect the remaining coin and nail to complete the circuit. Does the LED light up?

Make an electromagnet

You can make a temporary magnet by passing an electrical current through a coil of wire wrapped around a nail.

what you need

- ruler
- roll of thin electrical wire
- wire cutter
- long nail
- 2 crocodile clips
- battery or set of batteries in a battery holder
- switch
- metal paperclips

1 Measure out about 30 cm (12 in) of wire. Use the wire cutters to strip 2 cm (1 in) from each end of the wire.

2 Tightly wrap the wire around the nail, but leave the stripped ends uncoiled. The wire should go round the nail at least 50 times.

3 Attach a crocodile clip to each stripped end of the wire.

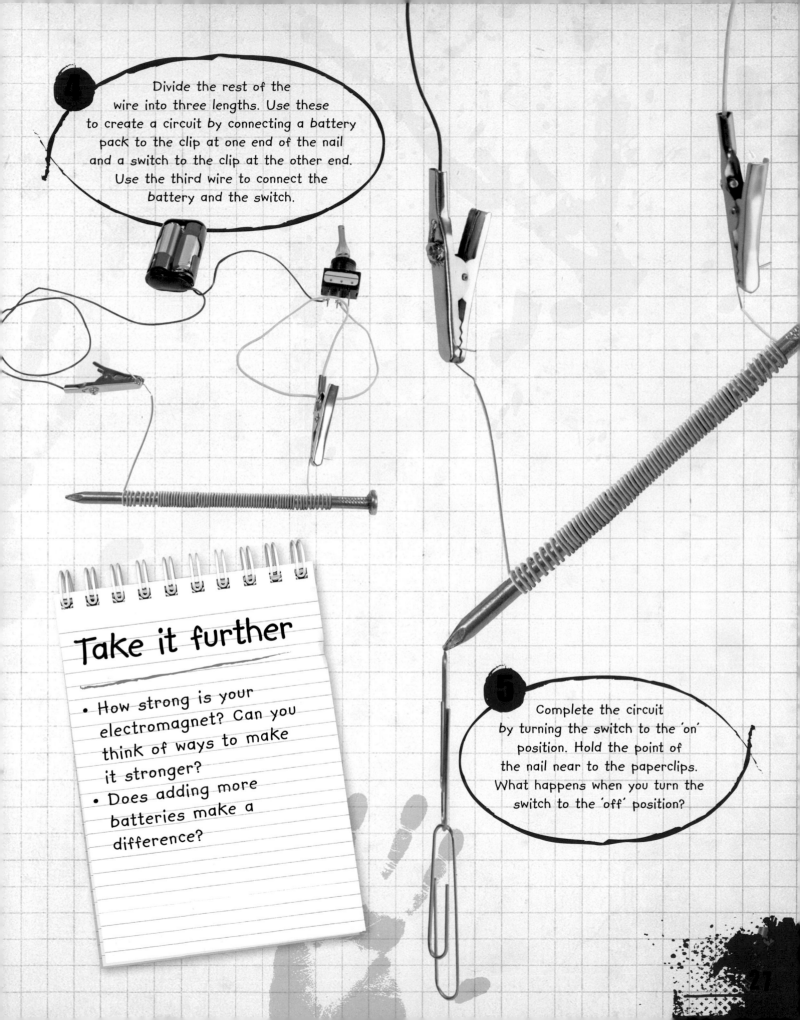

4 Divide the rest of the wire into three lengths. Use these to create a circuit by connecting a battery pack to the clip at one end of the nail and a switch to the clip at the other end. Use the third wire to connect the battery and the switch.

Take it further

- How strong is your electromagnet? Can you think of ways to make it stronger?
- Does adding more batteries make a difference?

5 Complete the circuit by turning the switch to the 'on' position. Hold the point of the nail near to the paperclips. What happens when you turn the switch to the 'off' position?

Make a fan

This cooling fan is powered by a motor, which turns electricity into a turning movement.

what you need

- pair of compasses
- ruler
- sharp pencil
- card
- scissors
- shoe box (for base)
- sticky tape
- electric motor
- batteries in a battery holder
- insulated wires
- switch
- modelling clay
- food can

24 cm (10 in)

1 To make the blades of the fan, use the compasses and ruler to draw a circle 24 cm (10 in) in diameter. Cut it out.

2 Use the sharp pencil to widen the hole made by the compasses. Cut four slits 10 cm (4 in) long from the outer edge in. Bend one side of each of the blades back a little.

3 Using the sticky tape, stick the motor to the top of the box.

4 Create a switch-operated circuit, connecting the batteries to the electric motor and a switch. Tape the switch to the inside of the box, with the switch poking through. Use the food can to stop your fan from falling over.

5 Attach the blades to the motor spindle using some modelling clay to hold them in place. Make sure there is a gap between the blades and the fan body so that the blades can spin freely.

6 Press the switch to 'on' to turn on the fan.

Take it further

- Bend the blades of your fan a little further. Does this make the breeze stronger?
- Add more batteries to your fan. Does the fan turn faster?

GLOSSARY

bass
The lowest pitch in music.

battery
A device containing an electric cell or a series of electric cells that changes chemical energy into electrical energy.

chemical reaction
A process in which one or more substances are changed into others.

circuit
A closed path through which an electric current flows or may flow.

circuit board
An insulated board, usually found in gadgets, such as computers, on which interconnected circuits and components, such as microchips, are mounted.

circuit diagram
A scientific drawing used to represent a circuit. Each component, such as a resistor, is represented by a specific symbol.

components
The parts of something. For example, a light bulb is a component in a circuit.

current
A flow of electric charge.

electromagnet
A device consisting of a coil of wire wrapped around an iron core that becomes magnetised when an electric current passes through the wire.

incandescent bulb
A light bulb that produces light using electricity to heat a thin wire, called a filament.

LCD (Liquid Crystal Display)
A low-power, flat-panel display used in many digital devices to display text or images.

LED (Light Emitting Diode)
An electric device that gives off light when an electric current passes through it.

mains power
The electricity that flows from a power station into homes or buildings.

microchip
A tiny slice of semiconducting material, on which an entire integrated circuit is formed.

power station
The place where mains electricity is generated.

rechargeable battery
A battery that can be used over and over again. When the battery goes flat, it is electrically recharged.

resistor
A device that is used to control current in an electric circuit by reducing the flow of electricity. Light bulbs and buzzers are both examples of resistors.

static electricity
Electricity that builds up on objects without flowing. Sometimes a static charge can become so large that it jumps, causing sparks.

switch
A component found in an electrical circuit that interrupts or diverts the current to break the circuit.

voltage
A measure of the difference in electric charge between two points, which drives an electric current between them. A high voltage can create a dangerous current.

TOPIC WEB

Use this topic web to discover themes and ideas in subjects that are related to electrical gadgets.

English
Think about your favourite gadget or the one you use most often. Write step-by-step instructions aimed at someone who has never used the gadget.

Geography
Use the Internet to find out more about how the world is linked – how have gadgets, such as mobile telephones and the computers, helped to do this?

Design and technology
Use your knowledge of electricity and materials to design and make your own torch.

Electrical gadgets

Citizenship
How do gadgets get the news to us? Write a news report on a specific event. Detail which gadgets you have used for research and which ones you have used to present your news item.

History
Make a timeline of a gadget that changed the world – how was the gadget used throughout history, and how has it changed over the years?

HOW THINGS WORK!

Contents of all titles in the series:

ELECTRICAL GADGETS

MUSICAL INSTRUMENTS

SENSORS

SIMPLE MECHANISMS

WAYLAND